ALL ABOUT BABY CHEETAHS

by Martha E. H. Rustad

PEBBLE
a capstone imprint

Pebble Emerge is published by Pebble, an imprint of Capstone.
1710 Roe Crest Drive
North Mankato, Minnesota 56003
www.capstonepub.com

Copyright © 2022 by Capstone. All rights reserved. No part of this publication may be reproduced in whole or in part, or stored in a retrieval system, or transmitted in any form or by any means, electronic, mechanical, photocopying, recording, or otherwise, without written permission of the publisher.

Library of Congress Cataloging-in-Publication Data
Names: Rustad, Martha E. H. (Martha Elizabeth Hillman), 1975- author.
Title: All about baby cheetahs / by Martha E.H. Rustad.
Description: North Mankato, Minnesota : Pebble, [2022] | Series: Oh baby! | Includes bibliographical references and index. | Audience: Ages 5-8 | Audience: Grades K-1 | Summary: "There's a new baby joining the family. It's a cheetah cub! Learn all about baby cheetahs, including what they eat, what they weigh, how they're raised, and how big they grow"—Provided by publisher.
Identifiers: LCCN 2021002813 | ISBN 9781663907790 (hardcover) | ISBN 9781663907769 (pdf) | ISBN 9781663907783 (kindle edition)
Subjects: LCSH: Cheetah—Infancy—Juvenile literature.
Classification: LCC QL737.C23 R875 2022 | DDC 599.75/91392—dc23
LC record available at https://lccn.loc.gov/2021002813

Image Credits
Newscom: Suzi Eszterhas/Minden Pictures, 12; Shutterstock: Alexey Osokin, 5, 6, Chris Fouri, 11, Eric Isselee, back cover, gd_project, 21 top, Louise Victor, 13, Maggy Meyer, 9, Manny DaCunha, 20 top, Mary Ann McDonald, 7, nwdph, 17, 19, oksana2010, 20 bottom, Paul Tessier, 14, Stu Porter, cover, 15, 18.

Editorial Credits
Editor: Alison Deering; Designer: Jennifer Bergstrom; Media Researcher: Tracy Cummins; Production Specialist: Tori Abraham

All internet sites appearing in back matter were available and accurate when this book was sent to press.

Table of Contents

Newborn Cheetahs 4
Sights and Sounds 8
Eating and Hunting 12
Growing Up! 16
    Run Like a Cheetah 20
    Glossary 22
    Read More 23
    Internet Sites 23
    Index 24

Words in **bold** are in the glossary.

NEWBORN CHEETAHS

Look! There are new babies on the African **savanna**. These babies are cheetah **cubs**. Cheetahs are part of the cat family.

Cubs grow inside the mother for three months. Mothers give birth to **litters** of three to five cubs. Each cub weighs about 1 pound (0.5 kilogram). The mother licks her cubs clean.

The cheetah mother hides her cubs in a nest in tall grasses. She keeps them hidden from **predators**. Every few days, she moves them to a new nest. She picks each cub up by its neck fur.

A cheetah mother raises her cubs by herself. Newborn cubs are blind and helpless. Their mother's strong eyesight looks for danger.

SIGHTS AND SOUNDS

Adult cheetahs are tan with black spots. But baby cheetahs have fuzzy silver fur on their backs. This helps keep them hidden in tall grasses. It also helps the cub stay cool in hot weather.

Cheetahs make lots of sounds. But they do not roar. They snarl, hiss, and grunt at enemies. Mother cheetahs purr.

Hungry cubs mew and squeak. They also chirp. Chirps help mothers and cubs find each other if they are apart.

EATING AND HUNTING

When they are born, baby cheetahs drink milk from their mother's body. By 6 weeks old, they have teeth.

A mother cheetah hunts **prey** daily for her cubs. She brings them meat. Cheetahs hunt animals like **antelope**. They also eat rabbits or other small **mammals**.

After about two months, cheetah cubs start to follow their mother. They learn to hunt. Cubs watch their mother closely. She sneaks up on prey. She creeps closer. Then she runs fast!

Cheetahs are the fastest animals on land. They can run up to 60 miles (97 kilometers) per hour. But they can only run that fast for 20 to 30 seconds.

When she catches prey, a mother cheetah calls to her cubs. Their tiny, sharp teeth rip and chew meat.

GROWING UP!

Cheetah cubs play with their brothers and sisters as they grow. They sneak up on each other. They **pounce** and chase. They practice hunting skills. Soon they will start to hunt small animals.

Young cheetahs need lots of hunting practice. At first, they cannot catch much. They still need help from their mother to survive.

Cheetahs are fully grown by age 2. They leave their mother. They are ready to take care of themselves.

Some adult cheetahs live alone. Male cheetahs often live with their brothers. Female cheetahs have cubs around age 3. Cheetahs live for about 12 years.

RUN LIKE A CHEETAH

Cheetahs can run very fast for about 30 seconds. In that time, they travel about 1,600 feet (500 meters). How far can you run in 30 seconds?

What You Need

- a ruler or tape measure
- tape or chalk
- a stopwatch

What You Do

1. Find a friend or adult to be your helper.
2. Mark a starting line. Ask your helper to time you.
3. When your helper says "go," run as fast as you can for 30 seconds.
4. Measure how far you ran.

How far did you run?

Did you run as far as a cheetah?

Glossary

antelope (AN-tuh-lohp)—an animal that looks like a large deer and runs very fast

cub (KUHB)—a young animal that eats meat

litter (LIT-er)—the young born to an animal at a single time

mammal (MAM-uhl)—a warm-blooded animal that breathes air; mammals have hair or fur; female mammals feed milk to their young

pounce (POUNS)—to jump on something suddenly and grab it

predator (PRED-uh-ter)—an animal that hunts other animals for food

prey (PRAY)—an animal that is hunted or killed by another animal for food

savanna (suh-VAN-uh)—open lands with grasses and short shrubs

Read More

Brandle, Marie. *Cheetahs*. Minneapolis: Jump!, Inc. 2021.

Jaycox, Jaclyn. *Cheetahs*. North Mankato, MN.: Capstone, 2021.

Murray, Julie. *Cheetahs*. Minneapolis: Abdo Publishing, 2020.

Internet Sites

Cheetah Conservation Fund: Kids
cheetah.org/kids/cheetah-facts/

Cheetah Cubs in the Wild
safeYouTube.net/w/JUJeb

San Diego Zoo: African Cheetah
kids.sandiegozoo.org/animals/african-cheetah

Index

adulthood, 18–19

African savanna, 4

appearance, 8

birth, 4

diet, 12–13, 15

eyesight, 7

habitat, 4, 6, 8

hunting, 13, 14–15, 16

lifespan, 19

litters, 4

playing, 16

predators, 6

prey, 13, 14–15, 16

siblings, 4, 16, 19

sounds, 10

speed, 14–15

weight, 4